Younes Sbihi
Abdelghani Iddar

Helicobacter pylori infection Pathogenicity and Diagnosis

Younes Sbihi
Abdelghani Iddar

Helicobacter pylori infection Pathogenicity and Diagnosis

Helicobacter pylori infection

ScienciaScripts

Imprint

Cover image: www.ingimage.com

This book is a translation from the original published under ISBN 978-3-8416-3606-5.

Publisher:
Sciencia Scripts
is a trademark of
Dodo Books Indian Ocean Ltd. and OmniScriptum S.R.L publishing group

120 High Road, East Finchley, London, N2 9ED, United Kingdom
Str. Armeneasca 28/1, office 1, Chisinau MD-2012, Republic of Moldova, Europe
Printed at: see last page
ISBN: 978-620-7-65818-3

Table of contents

Introduction

Helicobacter pylori is the most widespread infection worldwide, and is a bacterium that colonizes exclusively the gastric mucosa, playing an important role in the pathogenesis of gastroduodenal disorders such as gastritis, ulcer and gastric adenocarcinoma **[1-2]**. This bacterium was discovered by JR Warren and BJ Marshall in the early 1980s (Nobel Prize in Physiology and Medicine 2005). In 1994, the World Health Organization's International Agency for Research on Cancer (IARC) listed it as a type 1 carcinogen.

The involvement of *Helicobacter pylori* in various gastric pathologies has brought about a conceptual change, as this is the first time a bacterium has been considered responsible for a gastric process that has been treated in a palliative but not curative manner. Infection is acquired mainly during the first two decades of life, evolving over several years. It is the result of interactions between the bacterium's pathogenic factors, the host's inflammatory and immune responses, and its environment.

Helicobacter pylori infection is universally widespread, and can affect all populations, but is particularly prevalent where socio-economic conditions are unfavorable. Prevalence is 70-90% in developing countries, but only 20-30% in industrialized countries **[3]**. In France, the infection rate is currently 20-50% in the adult population **[4]**. In Morocco, a prevalence study published in 2013, carried out in patients presenting with various gastric pathologies, revealed the presence of *H. pylori* bacteria in 69% of patients, with a higher prevalence of infection (80.2%) in the 31-40 age group **[5].**

The mode of transmission of *H. pylori* is still uncertain. Given that *H. pylori* bacteria have been isolated from stools, saliva and dental plaque, this suggests that transmission is possible via the oral-oral or faeco-oral route **[6,7]**. Transmission is favoured by poor hygiene, unsanitary drinking water, poor food hygiene and promiscuity **[8]**.

This bacterium has certain characteristics that enable it to survive in the stomach and colonize the mucosa. In particular, *H. pylori* produces a urease that hydrolyzes the urea in gastric fluid into ammonia, allowing alkalinization of the environment and survival of the bacterium in the acid stomach. Its flagella give it great mobility in the mucus. The existence of an adhesin system enables the bacterium to bind to receptors on gastric epithelial cells. **[9-10]**

Diagnosis and treatment of *Helicobacter pylori* are recommended and indicated in subjects presenting symptoms of ulcer or dyspepsia. Diagnosis relies on a number of techniques of varying performance, the advantages and limitations of which must be clearly understood. It is based on a wide range of methods, grouped according to whether they are invasive or non-invasive **[11-12]**:

Non-invasive methods:

- *Serological tests*: based on the detection of *anti-H. pylori* IgG antibodies in serum. They appear 1 to 3 months after the onset of infection and persist throughout the course of the disease. These antibodies then gradually diminish. Serology is used to monitor and evaluate therapeutic efficacy. The efficacy of therapeutic eradication is accompanied by a significant reduction (50% in 6 months) in circulating immunoglobulins.

- *Detection of antibodies in saliva or urine*: The detection of antibodies in saliva or urine could be used in the doctor's office. However, the sensitivity and specificity of these tests are lower than those of serology. At present, they have no place in routine practice.

- *Detection of* H. pylori *antigens* in *stool* can **be** performed using an enzyme-linked immunosorbent assay (ELISA). The performance of this test looks promising, with a sensitivity of 94% and a specificity of 93% compared with conventional diagnostic tests (anatomopathology, urease test, culture). The non invasive nature of this test means it can be used both routinely and for epidemiological studies, particularly in children.

- *Labeled urea breath test:* This test is used to detect active infection. The principle is based on measurement of the quantity of CO_2 in exhaled air, usually labeled with the stable isotope carbon-13, following passage of labeled CO_2 into the bloodstream. The formation of CO_2 results from gastric hydrolysis by *H. pylori* urease of carbon13-enriched urea (C13) ingested by the patient. Exhaled air is collected at 0 and 30 minutes. Thus, by comparing the quantity of labelled CO_2 contained in the exhaled air, it is possible to prove whether or not the subject is infected with *H. pylori*. The sensitivity and specificity of this test are over 95%. False negatives may be due to treatment with antibiotics, proton pump inhibitors, bismuth salts or sucralfate. This implies stopping these types of treatment two weeks before the test. Despite data showing the relationship between the extent of CO_2 excretion and the extent of *H. pylori* infection, the urea breath test remains a qualitative test.

Invasive methods:

One or more biopsies are taken from the gastric mucosa to assess the presence of the bacterium by detecting its urease activity, visualizing histological staining or culturing it. Microbiological culture also enables antibiotic susceptibility testing. Over the last twenty years, bacterial genome detection techniques have been developed for the diagnosis and detection of resistance or pathogenicity genes in *H pylori* **[13-14]**.

Treatment is based on several antibiotics and a drug that reduces stomach acidity (called a proton pump inhibitor or PPI), and lasts 10 or 14 days, depending on the protocol. In 80-90% of cases, the infection is eliminated. In 10-20% of cases, treatment may fail, either because the bacteria are resistant to the antibiotics used, or because patients have difficulty following the treatment. If the first treatment has not eliminated the bacteria, the doctor will propose a new treatment with different antibiotics from the first, in order to overcome the problem of bacterial resistance. Because of the risk of treatment failure, it is essential to check that the infection has been eliminated.

At the end of the 2000s, probabilistic quadritherapies (PPI and three anti-infective agents) supplanted tritherapies, which retain their value when guided by prior determination of antibiotic sensitivity (imperative in the event of repeated failures) **[15]**.

History of the discovery of the bacterium

Microorganisms belonging to the Helicobacter genus are bacteria of great interest to medicine, not least because of their involvement in gastroduodenal pathology. More precisely, in 1906, Kreintz and his colleagues publicized the discovery of microorganisms, similar to those mentioned above, in certain patients with gastric cancer **[16]**.

The presence of spiral microorganisms was described by Doenges in 1938. The same author described these bacteria as dense with only two or three spirals and sometimes found inside gastric parietal cells. Two years later, in 1940, Freedberg and Barron detected the presence of curved bacteria in 40% of samples obtained from partial gastrectomies. In their book, these authors cite the association of these micro-organisms with gastric ulcers **[17]**.

Steer and Colin-Jones, in 1975, reported the presence of spiral-shaped Gram-negative micro-organisms in biopsies from patients with gastric ulcers, found beneath the gastric mucus layer and on the gastric mucosa itself **[18]**. Chronologically, however, the search for curved bacteria in gastric biopsies from Australian patients with gastroduodenal pathology began in 1980.

In 1983, the microbial genus was Campylobacter and the proposed species was pyloridis. One of Marshall and Warren's first scientific papers, published in The Lancet and entitled "Unidentified curved bacilli in the stomachs of patients with gastritis and peptic ulcer disease" **[19]**, was followed by a substantial body of work describing the discovery of *Campylobacter pyloridis* in the gastric mucosa of patients with gastritis and peptic ulcer disease.

In 1985, Marshall personally ingested these micro-organisms, called *Campylobacter pyloric*, in order to prove Koch's postulates. He produced a picture of gastritis with all the accompanying symptoms, as well as the pathological, microbiological and therapeutic data that were needed to demonstrate the pathogenic effect of these bacteria **[20]**. In 1987, Morris also established that ingestion of a dense inoculum of these bacteria leads to an image of gastric infection **[21]**.

These discoveries spread worldwide, and from 1983 onwards, the search for *Campylobacter pyloridis* began in gastric biopsy samples from patients. Following this research, the first publications appeared. Of the currently recognized species of the Helicobacter genus, pylori is the one of most interest, but it is necessary to refer to a whole range of species found in both humans and other animals.

The taxonomy *of H. pylori* is given in the following table:

Table 1. Taxonomy of Helicobacter pylori **(Marshall et al. 1985) [20].**

Kingdom	**Bacteria**
Division	Proteobacteria
Class	Epsilon Proteobacteria
Order	Campylobacter
Family	Helicobacteraceae
Type	Helicobacter
Species	*Helicobacter pylori*

Epidemiology

Gastric fluid, which is highly acidic, has a major bactericidal power. As a result, the stomach was classically considered sterile until the discovery of *H. pylori*, the only bacterium capable of surviving in such an environment. This bacterium is a good example of adaptation to a specific ecological niche, as certain characteristics enable it to survive in the stomach and colonize the mucosa.

Prevalence and incidence

Epidemiological studies have shown that *H. pylori* infection occurs worldwide (Figure 1) **[22]**.

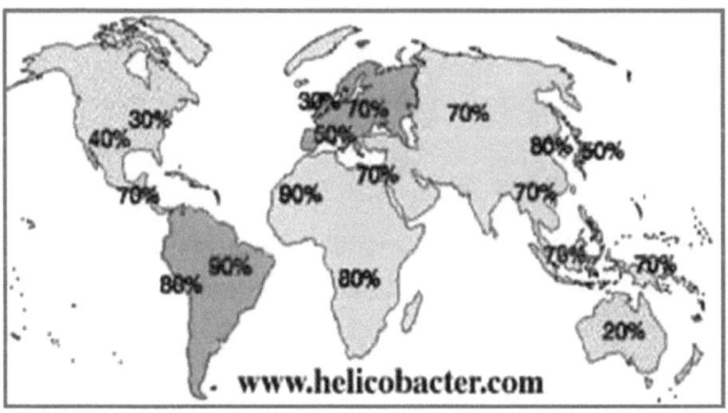

Figure 1: Prevalence of *H. pylori* infection in different regions of the world.

Helicobacter pylori is estimated to infect between 50 and 75% of the world's population, making it the most widespread infection worldwide **[23-24]**. Out of every 10 people infected with the bacterium, only one suffers from a disease and nine never develop it, the latter being a regulator of gastric homeostasis. Apart from age, the most important risk factor is socio-economic deprivation **[25]**. Many factors have been studied, but the common denominator is low economic status.

The incidence in children in industrialized countries is considered to be 1% per year, compared with 3 to 10% in developing countries **[26]**.

In developed countries, infection with this pathogen is rare in children, and increases progressively with age, reaching levels of 30% infestation by the age of 30, a value that remains constant at older ages **[24-25]**. In developing countries, most of the population is infected regardless of age, reaching levels close to 70%.

In Africa, overall prevalence is between 70% and 98% of the population **[27-28]**. In Morocco, *H.pylori* infection affects 70% of the population **[29-30]**. A retrospective cohort of 3,619 cases with gastroduodenal symptoms **[30]**, collected over a 5-year period, showed an overall prevalence of around 67.4%. This prevalence has been gradually declining since 1996. In Morocco, another prevalence study published in 2013, carried out in patients presenting with various gastric pathologies, revealed the presence of *H. pylori* bacteria in 69% of patients, with a higher prevalence of infection (80.2%) in the 31-40 age group **[5]**.

Transmission of infection

The mode of transmission of *H. pylori* is still uncertain. Humans are the main reservoir. Given that *H. pylori* bacteria have been isolated from stools, saliva and dental plaque, this suggests that transmission is possible via the oro-oral or faeco-oral route **[6-7]**. Transmission is favoured by poor hygiene, unsanitary drinking water, poor food hygiene and promiscuity **[23]**.

In developing countries, overcrowding, communal living conditions, lack of hygiene and frequent vomiting all contribute to contamination **[31]**. Certain cultural practices can also facilitate oral transmission.

In developed countries, oro-oral and gastro-oral contamination are also prevalent, particularly within the same family, via saliva and/or gastric fluid, from parents to children. Acquisition is also possible within couples, if one spouse is a carrier **[32]**.

Transmission via water sources and foodstuffs concerns developing countries, where the prevalence of infection is high and access to drinking water is limited. Consumption of raw vegetables has also been incriminated as a probable source of contaminated water **[33]**.

Helicobacter pylori infection is undoubtedly etiologically associated with the development of peptic, gastric or duodenal ulcers, and with the development of a particular type of gastric lymphoma (MALT lymphoma). It also plays a leading role in the development of gastric cancer; in fact, it is estimated to contribute 50% as an etiological agent in the multicausal chain of gastric cancer **[34-35]**. This bacterium has been classified as a type I carcinogen by the World Health Organization **[36]**.

Microbiological and biochemical characteristics

The discovery of *Helicobacter pylori* by Warren and Marshall in 1982 represented a milestone in the history of microbiology. A micro-organism appeared that was not only new to science, but led to a new concept in gastroduodenal pathology with far-reaching repercussions.

H. pylori is a Gram-negative bacillus, 2.5 µm to 5 µm long and 0.3 µm wide, with a helical or S-shaped spiral morphology. This bacterium has 4 to 6 flagella (compared with one flagellum in Campylobacters), giving it great mobility and enabling it to chronically colonize the gastric mucosa of half of humanity (Figure 2) **[37]**.

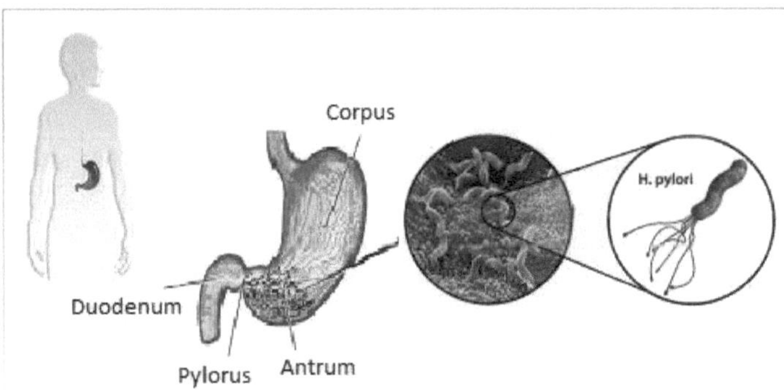

Figure 2: *Helicobacter pylori* affects the lower antrum of the stomach.

Gram appearance can be varied, with bacillary U-, C-, S- or comma-shaped forms. On aging cultures, bacillary forms give rise to round coccoid forms (Figure 3). These coccoids are unable to divide in culture, but maintain a certain metabolism and retain functional structures compatible with viability **[38]**.

H. pylori is a demanding bacterium, requiring long, difficult and delicate cultivation, and slow growth. Under optimal conditions, colonies appear in 3 to 7 days for primoculture and 2 to 4 days for subculture. Colonies are small, regular, round, domed, transparent and shiny.

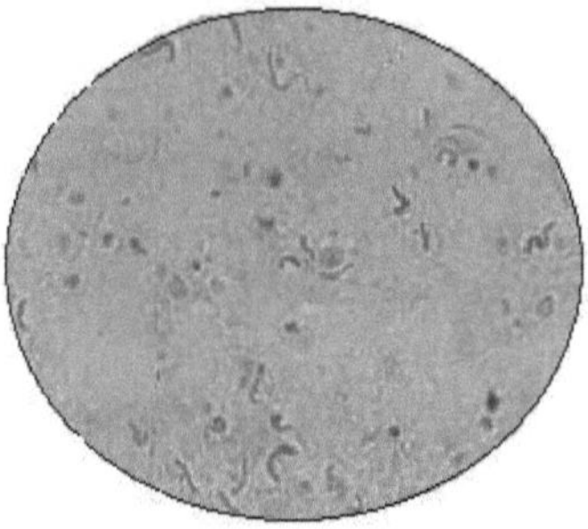

Figure 3: Microscopic examination of a culture after Gram staining. H. pylori appears in various forms: bacillary, curved, U-shaped, C-shaped or O-shaped.

Although it is a micro-aerophilic microorganism, *H. pylori* is capable of metabolizing glucose (the only carbohydrate it uses) by both oxidative and fermentative routes. In addition to the Entner-Doudoroff metabolic pathway, it possesses the pentose phosphate pathway and the tricarboxylic acid cycle. In terms of their metabolic requirements, all strains need the following amino acids: arginine, histidine, isoleucine, leucine, methionine, phenylalanine and valine **[39]**. They do not hydrolyze gelatin, starch, casein or tyrosine, and are negative for methyl red and Voges-Proskauer. This bacterium has oxidase, catalase and urease activity, the latter being its most important biochemical characteristic.

Urease is an enzyme that plays a major role in the pathogenesis of *H. pylori* infection. By hydrolyzing urea, normally present in the stomach, into ammonia, the bacterium buffers its own environment and protects itself from gastric acidity. This protection is essential to the bacterium, as without it, *H. pylori* is unable to survive at pH = 2. Catalase is an enzyme that catalyzes the breakdown of hydrogen peroxide into water and oxygen, protecting the bacterium from the harmful effects of oxygen metabolites, particularly oxygen peroxide. This catalase is not essential for bacterial growth and survival of *H. pylori* in vitro. However, catalase is considered a virulence factor, as catalase-negative mutants have never been found in the human stomach infected with *H. pylori* **[40]**.

Although *H. pylori* is highly homogeneous in terms of the biochemical characteristics used for diagnosis in clinical microbiology, it exhibits genetic variability through point mutations and inter- and intragenomic recombination, and colonization by various strains can occur. It is one of the most diverse species in the human biosphere **[41]**. The nucleotide sequence of the genome of two strains has been determined: *H. pylori* 26695 and J99, verifying the genomic variability of the strains **[42]**.

Modern analytical genomic methods, in particular 16S ribosomal ribonucleic acid (rRNA) sequence analysis, demonstrate that Campylobacter and Helicobacter are the principal members of a distinct group of bacteria (rRNA Superfamily VI) that is distantly related to other eubacteria **[42]**. Other members of the group are Arcobacter, recently excluded from the genus Campylobacter, and Wolinella, which consists of a single species, W. succinogenes, an inhabitant of the bovine rumen. The Helicobacter genus currently contains 10 officially named species, with others awaiting proper description and validation.

Pathologies and clinical aspects

Infection usually occurs in childhood, colonizing the gastric mucosa and producing a superficial gastritis that may remain for life, or be the starting point for the development of a duodenal ulcer or gastric neoplasia. Genetic and environmental factors, as well as the pathogenicity of the bacterium itself, influence the development of the disease in each individual **[43]**.

It is accepted that *H. pylori* infection is not only linked to peptic ulcer disease, gastric carcinoma and gastric MALT lymphoma **[44]**, but, without a clear understanding of the pathogenic mechanisms, it can cause iron-deficiency anemia, thrombocytopenia purpura or vitamin B12 deficiency.

Clinical aspects

Chronic gastritis

The term chronic gastritis indicates the presence of signs of inflammatory response of the gastric mucosa detectable on histological examination of gastric biopsy specimens, without symptoms or morphological abnormalities visible on endoscopy. The presence of *H. pylori* infection is manifested on the mucosal surface by an influx of neutrophils to the epithelium and a chorionic reaction in the form of an increased lymphoplasmacytic population **[45-46]**. Chronic gastritis can also be diffuse (pangastritis or multifocal gastritis).

Gastroduodenal ulcers

H. pylori infection is the leading cause of gastric and duodenal peptic ulcers. Between 15% and 20% of infected individuals develop a gastric or duodenal ulcer (Figure 4) **[47]**. The prevalence of gastric and duodenal ulcers has declined over the last few decades, but is increasing with increasing use of non-steroidal anti-inflammatory drugs **[48]**. The predominant symptom of ulcer disease is epigastric pain, which may be accompanied by fullness, bloating, early satiety and nausea **[49]**.

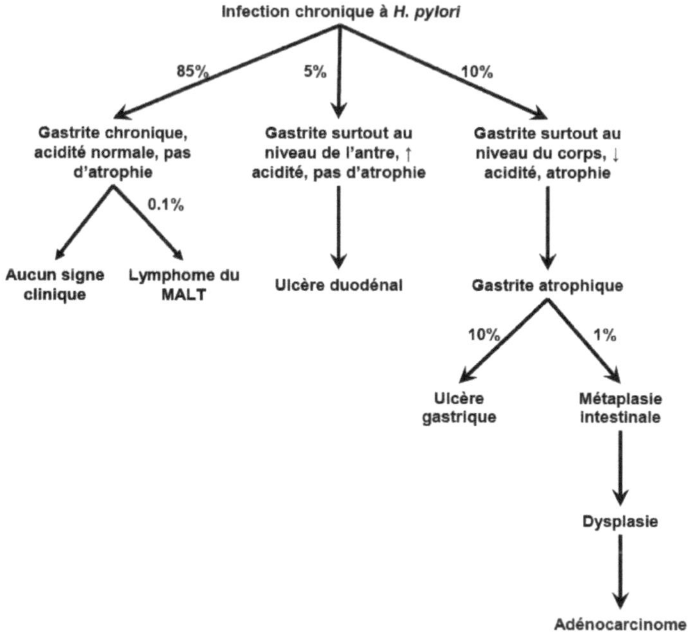

Figure 4. Gastrointestinal pathologies induced by *H. pylori* infection. In the vast majority of cases, *H. pylori-induced* gastritis remains asymptomatic. In around 0.1% of those infected, gastritis progresses to MALT lymphoma. In 5-10% of patients, gastritis progresses to gastric or duodenal ulcer. Finally, 1% of patients with atrophic gastritis will develop adenocarcinoma.

After eradication of the infection, recurrence of ulcer disease and its complications is virtually non-existent **[50]**.

Gastric mucosa-associated lymphoma (MALT)

Gastric MALT lymphoma ("Mucosa Associated Lymphoma Tissue") is a monoclonal proliferation of B lymphocytes in the marginal zone of lymphoid follicles.

In 90% of patients with low-grade gastric malignancy, *H. pylori* infection is detected, this microorganism being the main cause of these lymphomas. This causal relationship is reinforced by the histological regression of most early-stage lymphomas (75-85%) after eradication of the bacterium **[51]**.

At a more advanced stage of malignancy, cure by eradication of *H. pylori* is more uncertain, probably due to the accumulation of irreversible cellular genetic abnormalities **[52]**.

Gastric adenocarcinoma

H. pylori causes chronic inflammation of the gastric mucosa, which slowly progresses to the pre-neoplastic stages of gastric atrophy, intestinal metaplasia, dysplasia and gastric carcinoma.

The evolution of these histopathological lesions is known as Correa's cascade (Figure 5) **[46]**.

Figure 5. Different stages of the "Correa cascade" leading to adenocarcinoma intestinal type.

This infection is the main risk factor and most frequent cause of non-cardiac gastric carcinoma, the fifth most common malignancy, found in 90% of these patients **[35, 53]**. The risk of gastric cancer is up to 20 times higher in the infected population, and even higher in populations infected with the most aggressive bacterial strains.

Eradication of *H. pylori* reduces the risk of gastric cancer if it occurs before the onset of metaplasia **[34]**.

Pathogenicity mechanisms

There are a series of pathogenic factors or mechanisms that enable *H. pylori to* remain and damage the gastric mucosa once it comes into contact with the host.

Factors contributing to colonization of the gastric mucosa

H. pylori is able to survive and multiply in the gastric environment, which is unfavorable to the growth of most bacteria due to its acidic pH. The main characteristics that enable *H. pylori* to colonize gastric mucus are urease activity, microaerophilia, spiral shape, flagella and adhesins.

Urease

It is the most abundant enzyme produced by this bacterium, composed of two subunits, UreA and UreB, and regulates the pH around the bacterium **[54]**. Urease is a metalloenzyme that hydrolyzes urea into ammonium and carbon dioxide. The ammonia produced neutralizes the acidic environment, whereas an excessive increase in ammonia would be toxic to the bacteria. Regulation takes place via the Urel transporter. Ammonium plays an important role in the host immune response, activating monocytes and polymorphonuclear lymphocytes and inducing cytokine release, which contributes to gastric epithelial damage. It also produces a series of damages that affect microcirculation and epithelial cells, causing deep tissue necrosis, collaborating in the development of chronic atrophic gastritis and facilitating increased viral infections and carcinogenesis.

Antioxidant systems

Helicobacter pylori is a microaerophilic bacterium vulnerable to oxygen toxicity. The inflammatory response promoted by the bacterium during colonization gives rise to a large number of oxygen-reactive metabolites, so *H pylori* possesses the mechanisms to decode these metabolites (superoxide enzyme, catalase or peroxidase, peroxyrredoreductases, flavoprotein MdaB and the thioredoxin system), as well as repair the damage sustained (RecA, UvrABC proteins, endonuclease III, MsutS and RuvC) **[55]**.

Motility

H. pylori has 2 to 6 unipolar flagella, which give it mobility and enable it to penetrate the mucin layer. Flagella are mainly composed of the flagellins FlaA and FlaB **[56]**. These flagella are surrounded by a sheath similar in composition to that of the bacterial outer membrane, which provides protection against the acidic environment in which the bacteria evolve. In addition, this sheath reduces the release of immunogenic epitopes, explaining the absence of activation of the Toll Like Receptor 5 (TLR) innate immune pathway, which normally recognizes flagellins **[57]**. This could therefore contribute to the bacterium's persistence in the infected host. Indeed, non-flagellated strains of *H. pylori* only transiently colonize gnotobiotic piglets **[58]**.

The curved shape of *H. pylori* allows it to penetrate the mucin layer and reach the epithelial cells. *H. pylori* moves by chemoattraction to areas of the mucosa where the pH is higher **[59]**.

Outer membrane proteins and adhesion

The adhesion of *H. pylori* to the gastric mucosa is of great importance for both initial colonization and bacterial persistence. Different families of outer membrane proteins are involved in this process.

Bab-A (blood antigen binding adhesion): interacts with gastric epithelial cells via Lewis B antigens. Binding promotes a non-specific immune response and the development of antibodies, contributing to chronic gastritis, gastric ulcer development and gastric cancer **[60]**.

H. pyloriaA (*Helicobacter pylori* adhesin A): promotes binding of glycoconjugates to sialic acid on the surface of epithelial cells and neutrophils. It is encoded by the H. pyloriaA gene. It is an antigen recognized by antibodies and stimulates the proliferation of T and B lymphocytes **[61]**.

SabA (sialic acid binding adhesion): binds to sialic acid receptors on neutrophils, stimulating the production of reactive oxygen species **[62]**.

OipA (outer membrane inflammatory protein): although all strains possess the oipA gene, some do not express this adhesin. Its expression is associated with increased IL-8 production **[63]**. This protein is also associated with ulcers and constitutes a new virulence event **[64]**.

Pathogenicity factors contributing to gastric mucosal lesions

Numerous bacterial virulence factors are involved in the inflammation and tissue damage that can result from *H. pylori* infection. A distinction is made between factors conferring enhanced pro-inflammatory properties on the bacterium, principally the cag pathogenicity island (cag PAI), the OipA protein and the DupA protein, and factors acting directly on the host cell, of which the vacuolating cytotoxin (VacA) is the most extensively studied.

VacA vacuolizing cytotoxin

All *H. pylori* strains possess the VacA gene, which encodes a toxin in 50-60% of strains **[55]**. The VacA toxin, composed of two subunits p33 and p55, binds to host cells and induces vacuolization through the formation of selective anion channels. In addition, VacA appears to trigger the apoptotic cascade by directly affecting mitochondrial function and inducing cytochrome C release **[65]**. It can also activate the Fas/CD95 receptor associated with mitochondrial membrane rupture; increase the expression of cyclooxygenase 2, and thus vascular endothelial growth factor, associated with carcinomas; and disrupt the maturation of phagosomes into macrophages, allowing bacteria to survive there **[66]**.

There is considerable sequence diversity in the vacA genes of different strains of *H. pylori*, identifying polymorphism in three variable regions: signal sequence region, middle region and intermediate region **(figure 6) [55, 67]:** a 5' region, encoding the signal peptide (s) and the mature Nterminal protein (s1 or s2); an intermediate region (i), part encoding the p33 subunit (i1 or i2), and a middle region (m) encoding the p55 subunit involved in epithelial cell binding (m1 or m2). The s1/i1/m1 VacA type is highly active, while the s2/i2/m2 type is inactive. Intermediate forms are the most common. The s1/i1/m2 type is active on a smaller number of cell lines than the s1/i1/m1 type, as m2 binds more readily to cells. The s1/i2/m2 type is non-vacuolizing (mechanism unknown).

Cover et al. have shown that VacA-induced apoptosis is also dependent on the N-terminal part of the protein **[68]**. The VacA protein has also been described as being involved in blocking T cell proliferation: VacA is thought to induce a pathway involving a phosphorylation/activation cascade leading to activation of GTPase Rac and reorganization of the actin cytoskeleton **[69]**. These 2 phenomena could inhibit T cell activation and division **[70, 71]**. In the same context, a study showed that VacA could partially mimic the activity of an immunosuppressive drug by inducing local immune suppression **[72]**. In vivo, administration of the purified protein to mice led to degeneration of the gastric mucosa and recruitment of inflammatory cells **[73]**. Ultimately, this cytotoxin plays an active role in the colonization of the gastric mucosa by *H. pylori* **[74, 75]**.

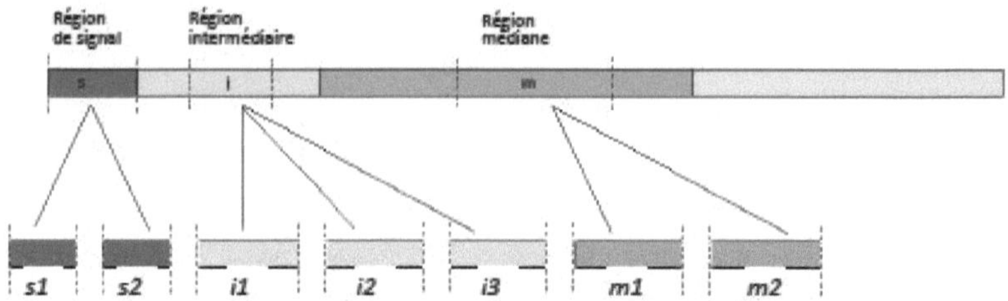

Figure 6. vacA gene structure and allelic diversity

The pathogenicity island cag

The cag pathogenicity island (cagPAI) is a gene cluster of approximately 40 kb that was acquired during evolution, through horizontal gene transfer from an unknown source. The cagPAI contains around 30 genes including cagA, and genes encoding a type IV secretion system (SST4) composed of several proteins **[76]**. The CagA protein is present in almost 90% of strains from Far Eastern countries such as Japan, Korea and China **[77]**. In contrast, only 60% of *H. pylori* strains isolated in Western countries (America, Africa and Europe) possess the gene encoding this protein. Far Eastern countries have a high incidence of gastric cancer, while Western countries have a lower incidence of gastric cancer. CagA-positive strains are associated with more pronounced inflammatory lesions of the gastric mucosa than CagA-negative strains, and thus increase the risk of gastric cancer **[78, 79]**.

Cytotoxin CagA

This is the most extensively studied virulence factor. It is located at the tip of the pathogenicity island and is considered a marker of the presence of the cagPAI island. This cytotoxin is associated with severe symptoms such as severe gastritis, mucosal atrophy, high risk of ulceration and gastric cancer. The most extensively studied CagA target is SH phosphatase. CagA varies in size between different strains, due to variation in a number of repeats of a carboxy-terminal amino acid sequence, known as the EPIYA motif. Differences in the EPIYA sequence (Glu-Pro-Ile-Tyr-Ala) indicate geographical and pathogenic variability **[80]**.

CagA appears to be an important virulence factor in gastric cancer, independent of the state of inflammation. For example, in a transgenic mouse model expressing CagA, a large number of B-cell lymphomas and advanced gastric adeno-carcinomas were observed without the presence of gastritis. Furthermore, eradication of *H. pylori* at the gastric atrophy stage did not significantly reduce the proportion of individuals who developed cancer over a 5-year period, implying that CagA was involved very early in the carcinogenesis process **[79]**. These data could call into question the gastritis-atrophy-metaplasia-carcinoma cascade initially described of the carcinogenesis process.

Other pathogenicity factors

Neutrophil-activating protein (NAP): This protein has a bacterioferritin function to capture free intracellular ferrous ions that can damage DNA. It can act as an adhesin when secreted or expressed on the bacterial surface.

Phospholipases: act as proteases, degrading the lipid-glycoprotein complex of the mucus layer covering the cells of the gastric epithelium.

Lipopolysaccharide (LPS): this is a low-toxicity virulence factor compared with other Gram-negative bacteria, explaining *H. pylori*'s ability to avoid provoking an effective host immune response and thus produce a chronic infection. It can modulate pepsinogen I activity in the stomach, affecting mucosal integrity **[81]**.

Other enzymes, such as mucinases, lipases, proteases and dismutases, enhance the bacterium's virulence and protect it from toxic metabolites.

Immune response

Colonization of the bacterium in the stomach can trigger a host immune response, resulting in systemic and local reactions involving neutrophilic infiltration and the appearance of *anti-H. pylori* antibodies. Gastritis is the result of a local host immune response and involves effusion of B and T lymphocytes, plasma cells, histiocytes and often polymorphonuclear cells **[82].**

Innate immunity

Gastric epithelial cells express molecular pattern recognition receptors (PRRs) on their surfaces, which interfere with *H. pylori* and activate expression. These include LTRs and Nod-Like Receptors (NLRs), which recognize pathogen-related receptors (PAMPs) such as LPS (Lipopolysaccharide), flagellins and peptidoglycan **[83]**. Interaction of the type IV secretion system (T4SS) with keratinocytes results in transfer of peptidoglycan into the cytoplasm, activation of Nod1 and expression of pro-inflammatory genes, as well as induction of secretion of the pro-inflammatory cytokines IL-8 and IL-12 through activation of NF- κB. Macrophages in the infected mucosa secrete pro-inflammatory cytokines and nitric oxide, and are potent bactericidal agents. *H. pylori* can survive phagocytosis by macrophages by inducing phagosome fusion to form megasomes without lysosomal fusion.

Adaptive immunity

H. pylori infection elicits a strong T cell response including both CD4+ (T helper, Th) and CD8+ (cytotoxic T) T cells, which contribute to inflammation **[82]**. With regard to CD4+ T lymphocytes, the main subpopulations induced by H. pylori infection are pro-inflammatory Th1 and Th17 and anti-inflammatory regulatory T cells (Treg). Th1 lymphocytes, favored by excessive IL-12 production by activated macrophages, produce pro-inflammatory cytokines (IFN-γ and TNF-α) which stimulate macrophages to secrete other pro-inflammatory cytokines and thus possess bactericidal activity **[84]**. Th17 lymphocytes secrete IL-17, IL-21 and IL-22, causing increased inflammation and PNN recruitment. The severity of gastritis is therefore related to the number of Th1 and Th17 lymphocytes **[85]**.

Humoral immunity

H. pylori stimulates the production of mucosal and systemic IgA and IgG antibodies, but the effect of antibodies on bacterial colonization remains controversial. One report showed that intragastric administration of specific monoclonal IgA induced protection against H. felis infection in mice **[86].** In contrast, others have shown that specific IgA and IgG in mice actually promote bacterial colonization and inhibit protective immune mechanisms **[86].** H. pylori is sensitive to the complement104-mediated bactericidal activity of serum, but B-cell-deficient mice can be successfully immunized **[87],** indicating that antibodies are not essential for protection. The B cell response plays an important role in pathogenesis by participating in an autoimmune process precipitated by H. pylori **[87]**. Here, antibodies cross-react with host antigens such as those of gastric epithelial cells and the parietal cell H+,K+-ATPase, potentially inducing inflammation and local damage.

The T cell response

In humans, the T-cell response to H. pylori is dominated by a T-helper 1 (Th1) response in the gastric mucosa **[82]**. Th1 cells produce IFNγ (Interferon gamma) and this type of response is associated with the expression of pro-inflammatory cytokines, e.g. TNF α (Tumor necrosis factor alpha), IL-12 and IL-18. When macrophages are activated in the presence of these type 1 cytokines, the resulting macrophages secrete pro-inflammatory factors with increased bactericidal activity compared to those activated in the presence of Th2 cytokines. The number of IFN γ-secreting cells in the infected human gastric mucosa correlates with the severity of gastritis **[88]**.

Outlook for future vaccines

The prevalence of antibiotic resistance among H. pylori isolates is increasing, and reports indicate that over 50% of isolates are resistant to metronidazole in parts of Asia and Africa **[89]**. Such antibiotic resistance is a problem for many pathogenic bacterial infections, and the best way to combat these infections on a large scale is probably through vaccination programs. Although vaccination seems the logical approach to combating H. pylori, vaccine research has not been straightforward. A number of vaccine antigens have been tested over the years, ranging from whole-cell lysates, and sub-infectious doses of live bacteria, to single or combined purified antigens, including urease, HP-NAP, CagA, VacA, catalase, heat shock proteins and lipopolysaccharides. The search for new, surface-exposed and highly conserved vaccine antigens among diverse strains worldwide is ongoing, using genomic and proteomic approaches **[90]**.

Diagnostic methods

Helicobacter pylori infection can be diagnosed by invasive methods, involving endoscopy with gastric biopsy, or by non-invasive methods. The performance of these techniques is varied, and requires a diagnostic strategy combining several of them to obtain optimum diagnostic performance. Some of these techniques provide only the notion of the presence or absence of *H. pylori* infection (standard serology, stool antigens, breath test, rapid urease activity). Others offer the possibility of assessing the consequences of the infection on the gastric mucosa (anatomopathology), establishing the antibiogram and typing the infecting strain (culture), and searching for certain resistance or virulence genes (culture, PCR on biopsy or stool) **[14]**. The choice of techniques to be used depends on the extent of the results required, and whether or not endoscopy is possible **[91, 92]**.

Diagnosis and treatment of *Helicobacter pylori* are recommended and indicated in subjects presenting with symptoms of ulcer or dyspepsia. The various invasive and non-invasive methods currently used for diagnosis are reported below **[11, 12]**.

Invasive methods

These methods usually require gastric biopsies taken during gastroduodenal endoscopy.

Rapid urease test

It is based on the presence of *H. pylori* urease. In conventional tests, the biopsy is placed in a tube containing urea and a pH indicator. The production of ammonia, thanks to the bacterium's high urease activity, causes the pH to rise and the phenol red used as a color indicator to turn pink.

There are also tests in which the reaction takes place on a test strip and the result is obtained within 1 hour, with similar sensitivity and specificity (90 and 95% respectively).

The result of urease tests depends on the bacterial load, requiring a minimum of 10^5 microorganisms per biopsy to obtain a positive result. In post-treatment controls, sensitivity decreases by up to 60% due to low bacterial load, the same occurs in patients with gastrointestinal bleeding due to peptic ulcer disease **[93]**.

Urease-positive bacteria contaminating the stomach can give false positives

Histological examination of biopsies

Histology allows direct visualization of bacteria in the tissue, as well as determination of histological damage, assessing the degree of gastritis, metaplasia or atrophy. Various stains such as Giemsa, hematoxylin-eosin or Warthin Starry with silver nitrate are used **[93]**. Three biopsies are recommended for histological sensitivity to exceed 90% **[94]**.

This technique does not distinguish between *H. pylori* and other species of the Helicobacter genus or bacteria of similar morphology **[95]**. Complementary techniques such as immunohistochemistry and in situ hybridization (FISH) achieve 98% sensitivity and 100% specificity **[96]**.

Bacteriological examination of biopsies

Culture is considered the reference method for diagnosing *H. pylori* , but this bacterium requires supplemented culture media to grow, a microaerophilic atmosphere and a prolonged incubation time: it can take between 5 and 7 days to obtain the strain. The best samples for isolation are biopsies taken from the antrum and the anterior and posterior bodies **[97]**. Biopsies are usually transported in semi-solid agar. They can be kept refrigerated for up to 24 hours. After that, they should be frozen at -70°C or in liquid nitrogen.

Culture sensitivity varies according to laboratory experience **[95]**. With the strain obtained, antimicrobial susceptibility studies can be carried out, as well as a better understanding of the bacterium, its mechanisms of pathogenicity and interactions with the host.

Identification of the *H. pylori* genus and species is straightforward. The culture requirements (microaerophilia, blood agar, selective medium) and the spiral appearance of a colony spread on a slide when observed microscopically after Gram staining, enable Helicobacter to be identified. The presence of strong catalase, oxidase and urease activity identifies the pylori species. Helicobacter non pylori do not grow under the conditions described. Only H. helmannii can infect the human stomach, but its corkscrew morphology is characteristic **[19, 98]**.

Culture is considered the reference method. It has the major advantage of enabling antibiotic treatment to be adapted to the sensitivity of the isolated strain, after antibiotic susceptibility testing. Various methods have been proposed for this purpose. The most widely used are the disk diffusion method and the determination of MIC (minimum inhibitory concentration) using epsilometers. In addition, strain culture is very useful for studying bacterial virulence markers.

Gene amplification (PCR)

This technique enables the detection of *H. pylori-specific* DNA sequences in gastric samples (biopsies, juice, mucus) or other samples (saliva, dental plaque, stool) **[95]**. However, this method does not confirm the presence of viable bacteria. Its value lies in the detection of bacteria present in low concentrations (post-therapeutic follow-up). PCR can also be used to detect chromosomal mutations in gastric biopsies, which are responsible for macrolide and fluoroquinolone resistance, and can be used as a typing tool in epidemiological analyses. The sensitivity and specificity of this technique are of the order of 95% **[99]**.

Currently, real-time PCR techniques are widely used. An attractive alternative technique is in situ hybridization of 16S rRNA gene sequences with a fluorescein-labeled probe (FISH: fluorescence in situ hybridization). Without gene amplification, but coupled with histology, this method offers greater specificity than conventional histology. The use of probes covering mutations involved in the acquisition of resistance to certain antibiotics also makes it possible to determine the sensitivity of strains, particularly to macrolides.

Despite all the problems associated with PCR (risk of contamination, need for qualified personnel and specialized equipment), this technique is proving to be a good alternative to culture for diagnosing *H. pylori* infection.

Non-invasive methods

Labeled urea breath test C^{13}

The labeled urea breath test is based on the demonstration of *Helicobacter pylori*'s urease activity in the stomach (see figure 6), which converts ingested urea into $^{13}CO_2$

and ammonia (NH_3) **[100]**. ^{13}C is a stable, non-radioactive isotope of carbon. It is completely harmless and can be used without special authorization.

There are several kits on the market worldwide, some of which are authorized and accepted in many countries: INFAI, TAU-Kit, UBTest, PYLORI CHEK, Héli-Kit...

The test generally proceeds as follows:

Before ingesting the labeled urea, the patient, fasting from the previous day and not having smoked, absorbs a citric acid solution which delays gastric emptying, thus increasing the contact time between Helicobacter pylori and the labeled urea;

he blows into a tube corresponding to the first collection of exhaled air (T0), which is useful because it may contain a certain proportion of ^{13}C, depending on the diet;

he then ingests urea labeled with a non-radioactive isotope (^{13}C) diluted in water, a dose that saturates the urease;

he remains at rest, sitting or lying down, without eating, drinking or smoking, for thirty minutes; he then blows into a second tube (T30).

The patient must fast for more than 6 hours, preferably overnight. The test takes about 40 minutes. If the test is to be repeated, it should not be done before the following day.

These samples are stable for long periods (5 weeks) **[100]**.

The air samples obtained are then analyzed using isotope ratio mass spectrometry (IRMS) or other technologies such as non-dispersive infrared spectrometry or laser-assisted analysis.

In the case of *Helicobacter pylori* infection, urea ingestion ^{13}C is followed by an enrichment of exhaled air with $^{13}CO_2$, its metabolite. The $^{13}CO_2$ produced diffuses into the epithelial cells and then into the circulating blood, and is eliminated by the lungs (figure 7) **[101]**. Analysis consists in determining the $C/^{1312}C$ ratio of exhaled air before and after ingestion of labeled urea **[102]**. The ammonia released by bacterial hydrolysis, as described below, is incorporated into the metabolism in the form of NH_4^+. This represents the most effective overall method for diagnosing and monitoring eradication.

In the absence of bacterial urease, the total amount of urea administered, after absorption by the gastrointestinal tract, will be metabolized as endogenous urea.

The sensitivity and specificity of this test are in excess of 90% **[3, 103]**, depending on the positivity threshold, usually set at 5 per thousand. However, false positives may occur due to the presence of other urease-positive bacteria. In addition, the attenuation of bacterial density by antibiotic or antisecretory treatment leads to false negatives.

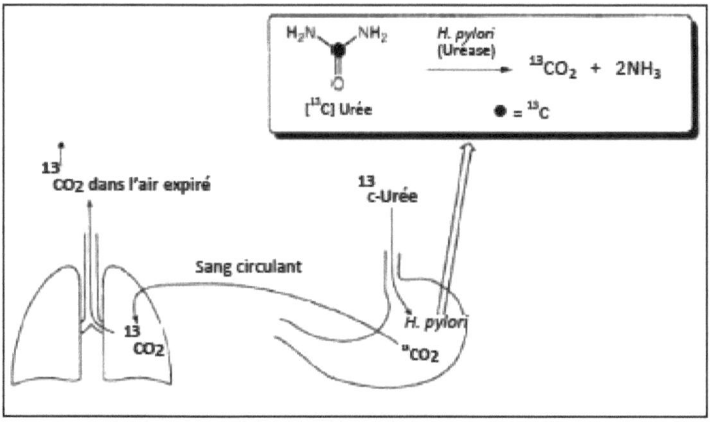

Figure 7. Principle of the Urea Breathing Test^{13} C (TRU^{13} C) for the detection of *H. pylori*.

All these methods, with the exception of serology, are less sensitive to recent antibiotic or antisecretory treatment. The use of certain antibiotics can reduce the bacterial mass and thus negate the tests. Tests should therefore be carried out at a distance, at least four weeks after antibiotic treatment, to monitor eradication. The increasingly frequent use of PPIs also reduces bacterial density and inactivates urease, without eradicating the bacteria; it is therefore important to discontinue PPI treatment at least two weeks before the breath test. If the patient is unable to stop antisecretory treatment, gastric dressings, which do not have these disadvantages, may be considered. Despite their lesser impact, a delay of seven days should be observed after taking H2 antihistamines **[3]**. In short, any investigation of infection should be carried out at least four weeks after taking antibiotics or antisecretory agents.

<u>Analysis by Isotope Ratio Mass Spectrometry (IRMS) :</u>

IRMS is a specialization of mass spectrometry, the principle of which is to separate ions according to their mass and charge. The sample to be examined is introduced into an ionization source. The sample molecules are then ionized (by electron bombardment) and accelerated. Detectors sort the ions according to their mass (mass 44 for $C\ O^{1216}{}_2$, mass 45 for $C\ O^{1316}{}_2$, etc.). These different ions travel different circular paths in the magnetic field, the radii of which depend on the mass-to-charge ratio. The intensity of the signals is thus specifically measured and reveals the$^{13}C:^{12}C$ ratio of the sample **[104, 105, 106].**

The result of the analysis of each sample is the delta (δ), expressed in parts per thousand (‰), which represents the relative difference between the isotopic ratio of the sample and that of a reference substance. The difference between the δ of samples (T = 30 and T = 0) from the same patient represents the delta over baseline (DOB). The final test result is generally considered positive when the DOB is greater than 5 ‰ **[103]**.

The $\delta^{13}C$ values expressed in relation to the reference standard are calculated using the following equation:

$$\delta^{13}{}_{Céch/réf}(‰) = 1000 \times (R_{éch} - R_{réf}) / R_{réf}$$

where

$R_{éch}$ is the isotopic ratio $C/^{1312}C$ of the test sample;
$R_{réf}$ is the isotopic ratio $C/^{1312}C$ of the reference standard.

Detection of bacterial antigens in stools

It involves the detection of *H. pylori* antigens in stools using polyclonal or monoclonal antibodies, the latter having similar efficacy to the urease breath test **[107]**. It is recommended as an alternative to TRU for the non-invasive diagnosis of *H. pylori* **[108]**.

Significant inter-test and batch-to-batch variability was observed. For faecal antigen testing, attention must be paid to the choice of reagents with the best diagnostic accuracy (sensitivity and specificity), as the evaluation showed that accuracy varied according to the reagents used and the conditions under which the stool was transported.

H. pylori-specific antigen testing is performed on stool, either fresh or stored at low temperature, or even frozen at -70°C. Various kits are available for this purpose. Various kits are available for this purpose; those using monoclonal antibodies offer the best results (Blanco et al, 2008). In practice, this test can be used for the primary diagnosis of infection, but is particularly useful for monitoring the efficacy of eradication treatment. Fairly practical in pediatrics, it could be used as an alternative to the respiratory test for monitoring eradication.

The need to collect and handle stools, then store the sample in a cool place until analysis, is an obstacle to the spread of this method.

Serological tests

Serology was one of the first methods used to diagnose *Helicobacter pylori* infection **[109]**. Infection with Helicobacter pylori leads to a strong, long-lasting immune response, including an increase in serum IgG levels, which remain high for the duration of the infection and then decline slowly, returning to levels comparable to those of uninfected subjects within 6 to 12 months.

The serum antibodies that appear during infection are specific to H pylori species antigens, so that satisfactory serodiagnostic results have been obtained even using crude bacterial extracts as antigens **[110]**. However, the possibility of cross-reactivity, particularly with certain antigens of the Campylobacter genus, has led to the use of purified antigens or antigen mixtures to demonstrate the antibody response for diagnostic purposes. Numerous kits corresponding to a variety of antigenic preparations are currently on the market. Most use an ELISA technique and detect IgG only; some also detect IgA.

Serological tests based on the determination of specific anti-Helicobacter IgG antibodies in blood need to be evaluated in each type of population before use. They do not always reflect an active infection, and cannot be used to assess the outcome of eradication at an early stage, as antibody levels, which are high during infection, only gradually decrease in the 4 to 6 months following the disappearance of the bacteria.

Because of its sensitivity, serology can be positive, testifying to a real infection, whereas direct tests, notably culture, can be negative for a variety of reasons. The lack of specificity attributed to serology reflects the inadequacy of the so-called reference tests with which it is compared. The current renewed interest in serology is linked to improvements in the serological tests available, particularly in terms of specificity, but also to the fact that in Western countries, more and more patients are receiving antisecretory treatment, leading to a reduction in the sensitivity of all diagnostic tests except serology **[111]**.

In addition, there are special clinical situations: digestive hemorrhage, gastric atrophy, gastric MALT lymphoma, gastric carcinoma, where serology has proved to be the most effective diagnostic test. Indeed, the small number of bacteria that colonize the stomach in these clinical circumstances reduces the sensitivity of direct diagnostic methods. Finally, recent studies have shown that serological tests may soon be able to detect bacterial markers associated with risks of severe evolution of *H. pylori* infection **[112, 113]**.

Rapid serological tests have been on the market for several years. These tests can be performed on blood, urine or saliva. They use immuno-chromatography or passive agglutination techniques. Their practical value is undeniable (rapidity, ease of use) but, until now, their insufficient sensitivity has meant that they have been reserved for epidemiological studies **[114]**.

Summary of diagnostic methods :

Non-invasive methods	Serology
	Breathing test
	Stool antigen testing
Invasive methods	Histological examination of biopsies
	Bacteriological examination of biopsies
	Gene amplification (PCR)
	Rapid urease test

Table 2. Summary of diagnostic methods for *Helicobacter pylori*

What to do in the event of suspected *H. pylori infection*

CNPHGE. 2017. Relevance of care - Diagnosis of Helicobacter pylori infection in adults. HAS May 2017.

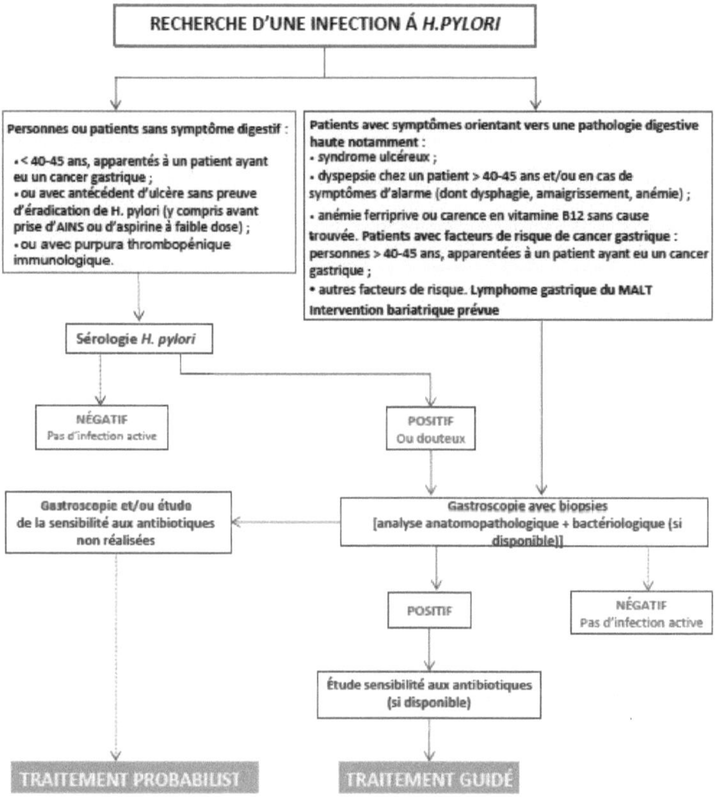

Outside the indications for gastroscopy

Non-invasive methods include serology, labeled urea breath test and stool antigen testing (CNPHGE. 2017. Relevance of care - Diagnosis of Helicobacter pylori infection in adults. HAS May 2017).

-Serology is indicated for H. pylori testing prior to treatment (choice of reagents with sensitivity/specificity ≥ 90%).

- The aim of serology is to avoid gastroscopy in H. pylori-negative patients, who are unlikely to have a severe lesion on gastroscopy, and to refer H. pylori-positive patients for gastroscopic evaluation. In the case of positive serology, gastroscopy is recommended to confirm the diagnosis of infection and, above all, to detect and manage any preneoplastic lesions induced by the bacterium; it also enables bacteriological examination and assessment of the bacterium's sensitivity to antibiotics, if this is feasible.

- Serology detects IgG antibodies. It is not indicated for monitoring eradication, as antibodies may persist for months or even years after the bacterium has been eradicated. There is no need to repeat a serological test.

The respiratory test is effective for pre-treatment diagnosis and eradication control, but is only reimbursed for eradication control.

Stool antigen testing is an effective tool for pre-treatment diagnosis and eradication control, but is not reimbursed.

Special situations

Serology is recommended in certain situations where other tests (labeled urea breath test, fecal antigen test, gastric biopsy examination) are less effective: bleeding ulcer, gastric atrophy, MALT lymphoma, use of antibiotics within the last 4 weeks, use of proton pump inhibitors (PPIs) within the last 2 weeks.

Treatment of H. pylori

Conventional treatment

The high prevalence of *H. pylori* infection, as well as its clinical implications, have always made it necessary to rationalize diagnosis of the infection and establish indications for its treatment, even though the latest Kyoto consensus considers gastritis due to *H. pylori to* be a serious disease, an entity in its own right, and its eradication is recommended in all cases **[115]**. However, indiscriminate eradication could lead to an increase in bacterial resistance, a high economic cost and the risk of eliminating strains that could have a beneficial effect on humans.

Tritherapy

The regimen includes a proton pump inhibitor (PPI) e.g. Omeprazole 40 mg, Amoxicillin (1g) and another antibiotic such as Clarithromycin or Metronidazole twice daily for 14 days **[116]**. Tritherapies are currently outdated as an experimental treatment, but may remain an excellent choice based on the sensitivity of the patient's immune system.

Quadritherapy

These 10- to 14-day probabilistic therapies combine either a PPI with 3 antibiotics or a PPI with Bismuth subcitrate and 2 antibiotics (Tetracycline and Metronidazole) **[117]**.

Sequential therapy

This therapy, presented by several Italian researchers, is based on the combination of a PPI and Amoxicillin (1g/2 days) for 5 days, followed by a PPI and Clarithromycin/Metronidazole (500g/2 per day) for 5 days **[118]**. Sequential therapy is currently considered an obsolete regimen that has been replaced by concomitant treatment.

Concomitant therapy

Used instead of sequential therapy where resistance to Clarithromycin is greater than 20% and Bismuth-based quadruple therapy is not available, this therapy involves the administration of 3 antibiotics (Metronidazole, Clarithromycin and Amoxicillin) and a PPI for 10 to 14 days, and is effective compared to conventional triple therapy **[119]**.In the presence of resistance to Clarithromycin or Metronidazole, but ineffective in the presence of double resistance to Clarithromycin and Metronidazole **[119]**

Factors influencing treatment efficacy

None of the therapies can improve H. pylori infection in 100% of patients, as several factors such as antibiotic resistance, poor treatment monitoring, higher bacterial load, low gastric pH and other factors can influence the outcome of eradication therapy **[120]**.

The most significant cause of treatment failure appears to be drug resistance, particularly to Clarithromycin, which reduces treatment efficacy by around 70% **[120]**. In contrast, H.pylori resistance to Amoxicillin and Tetracycline is low, accounting for less than 1% of isolates worldwide. Primary or acquired resistance of H.pylori to bismuth salts has never been reported.

Alternative treatment

Although mankind has always used plants as remedies, the synthetic molecules developed by pharmacopoeia have proved highly successful in the treatment of many pathologies. The current craze for phytotherapy stems from the innumerable side-effects associated with the use of these drugs, and the fact that many ailments remain uncured. The abundance of the plant kingdom on earth provides an invaluable source of active molecules for many ailments, and in the right quantities, plants offer a broad spectrum of new natural compounds with no side effects. For the past hundred years, the pharmaceutical industry has been involved in research into the therapeutic properties of medicinal plants, with the aim of identifying new active molecules and understanding their modes of action. The results of studies carried out on extracts of several types of medicinal plants have demonstrated antioxidant, anti-inflammatory, analgesic, anti-cancer and other effects. **[121]**.

Table 3 shows the plants most commonly used to treat H. pylori.

Table 3. Examples of plants used against *H. pylori*

Plant name	Reference	Description	Figure
Liquorice	Wan (2009) Wittschie (2009)	a perennial plant native to southeastern Europe and western Asia, which grows spontaneously in meadows. Realization root is known to have been used by our ancestors to treat digestive disorders (stomach ache, colitis).	
Ginger	Larousse encyclopedia of medicinal plants. (2001) Perotto (2013)	a tropical plant native to Asia that grows in wetlands. It has a large, fleshy rhizome that is used to treat dyspepsia, intestinal gas and colic.	
Lemon balm	Ibarra et al (2010) Soulimani et al (1991)	is a perennial herbaceous plant 40 to 80 cm tall with numerous square-section stems bearing broad leaves that are used in infusions to reduce abdominal pain and anxiety with digestive somatization.	
White marrubus	Ozenda (2014) Kaabeche (1990)	is a hairy herbaceous perennial with stems 30-60cm high, found in European and North African countries (Algeria, Tunisia, Morocco), growing in uncultivated areas, arid terrain and low-altitude mountainous regions.	

References

1. Hooi JKY, Lai WY, Ng WK, Suen MMY, Underwood FE, Tanyingoh D, Malfertheiner P, Graham DY, Wong VWS, Wu JCY, Chan FKL, Sung JJY, Kaplan GG and Ng SC. Global prevalence of helicobacter pylori infection: systematic review and meta-analysis. Gastroenterology 2017; 153: 420-429.. doi: 10.1053/j.gastro.2017.04.022.

2. Padda J, Khalid K, Cooper AC, Jean-Charles G. Association between Helicobacter pylori and Gastric Carcinoma. Cureus. 2021 May 22;13(5):e15165. doi: 10.7759/cureus.15165.

3. Bouyssou C. 2014. Helicobacter pylori: the essentials to understand. Actualités pharmaceutiques no. 536: 20-24. http://dx.doi.org/10.1016/j.actpha.2014.02.024

4. French Health Products Safety Agency. Prise en charge thérapeutique de l'éradication de Helicobacter pylori chez l'adulte et l'enfant. Saint-Denis: AFSSAPS; 2005

5. Essadik A, Benomar H, Rafik I, Hamza M, Guemouri L, Kettani A, Maachi F. Epidemiological and clinical aspects of Helicobacter pylori infection through a Moroccan study. Hegel Vol. 3 N° 3 - 2013 p 163-169. DOI : 10.4267/2042/51450

6. Everhart JE. Recent developments in the epidemiology of Helicobacter pylori. Gastroenterology Clinics of North America, 2000, 29 (3):78-559.

7. Haute Autorité de Santé. Recommendations en Santé Publique : Dépistage de l'infection à Helicobacter pylori. April 2010, France.

8. World Gastroenterology Organisation. 2011. World Gastroenterology Organisation Global Guideline: Helicobacter pylori in developing countries. J Clin Gastroenterol. 2011 May-Jun;45(5):383-8. doi: 10.1097/MCG.0b013e31820fb8f6..

9. McColl, K. E.: Helicobacter pylori infection. N Engl J Med 362, 1597-1604 (2010).

10. Hsu, P.-I., Yamaoka, Y., Goh, K.-L., Manfredi, M., Wu, D.-C., Mahachai, V.: Helicobacter pylori infection. BioMed Res Int 2015, 278308 (2015).

11. Mégraud Francis, Emilie Besséde, and Philippe Lehours. Diagnosis of Helicobacter pylori Infection. Helicobacter 19 (Suppl. 1): 6-10, 2014. doi: 10.1111/hel.12161

12. HAS. Conseil National Professionnel d'Hépato-Gastroentérologie. Pertinence des Soins, Diagnosis of Helicobacter pylori infection in adults. May 2017

13. Cambau E, Allerheiligen V, Coulon C, et al. 2009. Evaluation of a new test, genotype HelicoDR, for molecular detection of antibiotic resistance in Helicobacter pylori. J Clin Microbiol 2009; 47: 3600-3607.

14. Tamayo, E., Montes, M., Fernández-Reyes, M., Lizasoain, J., Ibarra, B., Mendarte, U., Pérez-Trallero, E. (2017). Clarithromycin resistance in Helicobacter pylori and its molecular determinants in Northern Spain, 2013-2015. Journal of Global Antimicrobial Resistance, 9, 43-46. doi:10.1016/j.jgar.2016.12.019

15. HAS. Fiche Pertinence des soins sur le traitement de l'infection par Helicobacter pylori chez l'adulte, May 2017. https://www.has-sante.fr/upload/docs/application/pdf/2017-06/dir83/helicobacter_fiche_pertinence_traitement.pdf.

16. Kreintz W. Ueber das auftreten von spirochaeten verschiendner form im mageninhalt bei carcinoma ventriculi. Dtsch Med Wochenschr 1906;32:872.

17. Freedberg AS, Barron LE. The presence of spirochetes in human gastric mucosa. Am J Dig Dis 1940;7:443-538.

18. Steer HW, Colin-Jones GG. Mucosal changes in gastric ulceration and their response to carbenoxolone sodium. Gut 1975;16:590-597.

19. Marshall BJ, Warren JR. Unidentified curved bacilli in the stomach of patients with gastrtis and peptic ulceration. Lancet 1984;1:1311-1315.

20. Marshall BJ, Amstrong JA, McGechie DB, Glancy RJ. Attempt to fulfill Koch's postulates for pyloric Campylobacter. Med J Aust 1985;142:436-469.

21. Morris a, Nicholson G. Ingestion of Campylobacter pyloridis causes gastritis and raised fasting gastric pH. Am J Gastroenterol 1987;87:192-198.

22. Cave DR. 1997. Epidemiology and Transmission of Helicobacter pylori Infection. How Is Helicobacter pylori Transmitted? Gastroenterology. 113: S9- S14.

23. Hunt R H, Xiao S D, Megraud F, Leon-Barua R, Bazzoli F, van der Merwe S, Vaz Coelho L G, Fock M, Fedail S, Cohen H, Malfertheiner P, Vakil N, Hamid S, Goh K L, Konturek B C Y, Krabshuis J, Le Mair A, World Gastroenterology Organization. Helicobacter pylori in developing countries. World Gastroenterology Organisation Global Guideline. J Gastrointestin Liver Dis. 2011 Sep;20(3):299-304.

24. Molaoa SZ. Prevalence of Helicobacter pylori infection and the incidence of the associated malignant and peptic ulcer disease (PUD) at Nelson Mandela Academic Hospital: a retrospective analysis. J Drug Assess. 2021 Feb 8;10(1):57-61. doi: 10.1080/21556660.2020.1854560.

25. Yuan C, Adeloye D, Luk TT, et al. The global prevalence of and factors associated with Helicobacter pylori infection in children: a systematic review and meta-analysis. Lancet Child Adolesc Health 2022; 6: 185-94.

26. Sherif, M., Mohran, Z., Fathy, H., Rockabrand, D.M., Rozmajzi, P.J., and Frenck, R.W. Universal high-level primary metronidazole resistance in Helicobacter pylori isolated from children in Egypt. J Clin Microbiol. 2004.42: 4832-4.

27. Secka, O., Berg, D.E., Antonio, M., Corrah, T., Tapgun, M., Walton, R., Thomas, V., Galano, J.J., Sancho, J., Adegbola, R.A., and Thomas, J.E. Antimicrobial susceptibility and resistance patterns among Helicobacter pylori strains from The Gambia, West Africa. Antimicrobial Agents Chem. 2013.57: 1231-7.

28. Oyedeji, K.S., Smith, S.I., Coker, A.O., and Arigbabu, A.O. Antibiotic susceptibility patterns in Helicobacter pylori strains from patients with upper gastrointestinal pathology in western Nigeria. Br J Biomed Sci. 2009.66: 10-3.

29. Lahbabi, M., Alaoui , S ., El Khazi, K., El Abkari, M., Nejjari, C., Amarti, A., Bennani, B., Mahmoud, M., Ibrahimi, A., and Benajah, D.A.Sequential therapy versus standard triple-drug therapy for Helicobacter pylori eradication. result of the HPFEZ randomized study. Clin Res Hepatol Gastroenterol. 2013. 37: 416-21.

30. Attaf N., Cherkaoui N., Choulli M.K., Ghazali L., Mokhtari A., SoulaymaniA. Epidemiological profile of helicobacter pylori infection in the Gharb-chrarda-Béni Hssen region. Biology and Health. 2004;4 (1):25-34.

31. Bardhan P.K. Epidemiological features of Helicobacter pylori infection in developing countries. Clin. Infect. Dis. 1997.25: 973-978.

32. Mapstone N.P, Lynch D.A.F, Lewis F.A, Axon A.T.R, Tompkins D.S, Dixon M.F,Quirke P.PCR identification of Helicobacter pylori in faeces from gastritis patients. Lancet ; 1993.341, 447.

33. Lu, Y., et al, Isolation and genotyping of Helicobacter pylori from untreated municipal wastewater.Appl Environ Microbiol, 2002. 68(3): p. 1436-9.

34. Chen HN, Wang Z, Li X, Zhou ZG. Helicobacter pylori eradication cannot reduce the risk of gastric cancer in patients with intestinal metaplasia and dysplasia: evidence from a meta-analysis. Gastric Cancer (2016) 19:166-175. DOI 10.1007/s10120-015-0462-7.

35. Moss SF. The Clinical Evidence Linking Helicobacter pylori to Gastric Cancer. Cell Mol Gastroenterol, 2016 Dec 27;3(2):183-191. doi: 10.1016/j.jcmgh.2016.12.001.

36. Asghar RJ, Parsonnet J: Helicobacter pylori and risk for gastric adenocarcinoma. Sem Gastroenterology 2001;12:203-208.

37. De Korwin JD. Helicobacter pylori 30 years after: What's new? Rev Med Interne. 2014 Sep;35(9):561-4.

38. Sarem Muhannad, Corti Rodolfo. Role of Helicobacter pylori coccoid forms in infection and recrudescence. Gastroenterol Hepatol. 2016 Jan;39(1):28-35. doi: 10.1016/j.gastrohep.2015.04.009.

39. Marais A, Mendz GL, Hazell SL, Mégraud F. Metabolism and genetics of Helicobacter pylori: the genome era. Microbiol Mol Biol Rev. 1999 Sep;63(3):642-74. doi: 10.1128/MMBR.63.3.642-674.1999.

40. De Korwin JD. [New recommendations for the diagnosis and the treatment of Helicobacter pylori infection].Presse Med. 2013.42(3):309-17. doi: 10.1016/j.lpm.2012.07.014.

41. Israel DA, Salama N, Krishna U, Rieger UM, Atherton JC, S Falkow S, R M Peek RM. Helicobacter pylori genetic diversity within the gastric niche of a single human host Proceedings of the National Academy of Sciences Dec 2001, 98 (25) 14625-14630; DOI: 10.1073/pnas.251551698.

42. Alm, R., Ling, LS., Moir, D. et al. Genomic-sequence comparison of two unrelated isolates of the human gastric pathogen Helicobacter pylori. Nature 397, 176-180 (1999). https://doi.org/10.1038/16495.

43. Mishra, S. Is Helicobacter pylori good or bad?. Eur J Clin Microbiol Infect Dis 32, 301-304 (2013). https://doi.org/10.1007/s10096-012-1773-9.

44. Salar, A. (2019). "Gastric MALT lymphoma and Helicobacter pylori." Med Clin (Barc) 152(2): 65- 71.

45. Versalovic J. Helicobacter pylori. Pathology and diagnostic strategies. Am J Clin Pathol. 2003;119:403-12. doi: 10.1309/5DTF5HT7NPLNA6J5.

46. Correa P, Piazuelo MB. The gastric precancerous cascade. J Dig Dis. 2012 13(1):2-9. doi: 10.1111/j.1751-2980.2011.00550.x.

47. Molaoa SZ. Prevalence of Helicobacter pylori infection and the incidence of the associated malignant and peptic ulcer disease (PUD) at Nelson Mandela Academic

Hospital: a retrospective analysis. J Drug Assess. 2021 Feb 8;10(1):57-61. doi: 10.1080/21556660.2020.1854560.

48. Musumba C., Jorgensen A., Sutton L., Van Eker D., Moorcroft J., Hopkins M., Pritchard D. M. and Pirmohamed M. (2012). The relative contribution of NSAIDs and Helicobacter pylori to the aetiology of endoscopically-diagnosed peptic ulcer disease: observations from a tertiary referral hospital in the UK between 2005 and 2010. Aliment. Pharmacol. Ther. 36, 48-56. 10.1111/j.1365-2036.2012.05118.x.

49. Malfertheiner P, Sipponen P, Naumann M, Moayyedi P, Megraud F, Xiao SD, Sugano K. and Nyren O. 2005. Helicobacter pylori eradication has the potential to prevent gastric cancer: a state of the art critique. Am J Gastroenterol, 100, 2100-2115.

50. Jin X, Li Y. 2007. Systematic Review and Meta-analysis from Chinese Literature: the Association Between Helicobacter pylori Eradication and Improvement of Functional Dyspepsia. Helicobacter 12(5) p. 541-546. https://doi.org/10.1111/j.1523-5378.2007.00520.x

51. Kawanaka, M., Watari, J., Kamiya, N. et al. Effects of Helicobacter pylori eradication on the development of metachronous gastric cancer after endoscopic treatment: analysis of molecular alterations by a randomised controlled trial. Br J Cancer 114, 21-29 (2016). https://doi.org/10.1038/bjc.2015.418

52. Du, M.Q. (2011) MALT lymphoma: many roads lead to NF-κB activation. Histopathology, 58, 26-38.

53. Padda J, Khalid K, Cooper AC, Jean-Charles G. Association between Helicobacter pylori and Gastric Carcinoma. Cureus. 2021 May 22;13(5):e15165. doi: 10.7759/cureus.15165.

54. Mobley, H.L., Island, M.D. and Hausinger, R.P. (1995) Molecular biology of microbial ureases. Microbiol Rev, 59, 451-480.

55. Cervantes-García E. Helicobacter pylori: mecanismos de patogenicidad. Rev Mex Patol Clin Med Lab. 2016; 63(2):100-109.

56. Eaton, K.A., Morgan, D.R. and Krakowka, S. (1992) Motility as a factor in the colonisation of gnotobiotic piglets by Helicobacter pylori. J Med Microbiol, 37, 123- 127.

57. Gewirtz, A.T., Yu, Y., Krishna, U.S., Israel, D.A., Lyons, S.L. and Peek, R.M., Jr. (2004) Helicobacter pylori flagellin evades toll-like receptor 5-mediated innate immunity. J Infect Dis, 189, 1914-1920.

58. Eaton, K.A., Suerbaum, S., Josenhans, C. and Krakowka, S. (1996) Colonization of gnotobiotic piglets by Helicobacter pylori deficient in two flagellin genes. Infect Immun, 64, 2445-2448.

59. Yoshiyama, H., Nakamura, H., Kimoto, M., Okita, K. and Nakazawa, T. (1999) Chemotaxis and motility of Helicobacter pylori in a viscous environment. J Gastroenterol, 34 Suppl 11, 18-23.

60. Olfat Farzad O., Zheng Quing, Oleastro Monica, Voland Petra, Borén Thomas, Karttunen Ruta, Engstrand Lars, Rad Roland Prinz, Christian, Gerhard Markus. Correlation of the Helicobacter pylori adherence factor BabA with duodenal ulcer disease in four European countries. FEMS Immunology & Medical Microbiology, Volume 44, Issue 2, May 2005, Pages 151-156, https://doi.org/10.1016/j.femsim.2004.10.010

61. Nyström J, Svennerholm A. Oral immunization with HpaA affords therapeutic protective immunity against H. pylori that is reflected by specific mucosal immune responses. Vaccine . 2007 Mar 30;25(14):2591-8. doi: 10.1016/j.vaccine.2006.12.026

62. Yamaoka Y. Mechanisms of disease: Helicobacter pylori virulence factors. Nat Rev Gastroenterol Hepatol . 2010 Nov;7(11):629-41. doi: 10.1038/nrgastro.2010.154

63. Yamaoka Yoshio, Kwon Dong H., and Graham David Y.. A Mr 34,000 proinflammatory outer membrane protein (oipA) of Helicobacter pylori. Proc Natl Acad Sci U S A. 2000 Jun 20; 97(13): 7533-7538. doi: 10.1073/pnas.130079797

64. Oleastro, M., Cordeiro, R., Ferrand, J., Nunes, B., Lehours, P., Carvalho-Oliveira, I., Mendes, A.I., Penque, D., Monteiro, L., Mégraud, F. and Ménard, A. (2008) Evaluation of the clinical significance of homB, a novel candidate marker of Helicobacter pylori strains associated with peptic ulcer disease. J Infect Dis, 198,1379-1387.

65. Palframan SL, Kwok T and Gabriel K. Vacuolating cytotoxin A (VacA), a key toxin for Helicobacter pylori pathogenesis. Front. Cell. Infect. Microbiol., 12 July 2012 | https://doi.org/10.3389/fcimb.2012.00092

66. Cover TL and Blanke SR. Helicobacter pylori VacA, a paradigm for toxin multifunctionality. Nat Rev Microbiol . 2005 Apr;3(4):320-32. doi: 10.1038/nrmicro1095.

67. Rhead JL, Letley DP, Mohammadi M, Hussein N, Mohagheghi MA, Hosseini ME, Atherton JC. A new Helicobacter pylori vacuolating cytotoxin determinant, the intermediate region, is associated with gastric cancer. Gastroenterology . 2007 Sep;133(3):926-36. doi: 10.1053/j.gastro.2007.06.056.

68. Sharndama HC, Mba IE. Helicobacter pylori: an up-to-date overview on the virulence and pathogenesis mechanisms. Braz J Microbiol. 2022 Mar; 53(1):33-50. doi: 10.1007/s42770-021-00675-0.

69. Hofman, P., Waidner, B., Hofman, V., Bereswill, S., Brest, P. and Kist, M. (2004) Pathogenesis of Helicobacter pylori infection. Helicobacter, 9 Suppl 1, 15-22.

70. Boncristiano, M., Paccani, S.R., Barone, S., Ulivieri, C., Patrussi, L., Ilver, D., Amedei, A., D'Elios, M.M., Telford, J.L. and Baldari, C.T. (2003) The Helicobacter pylori vacuolating toxin inhibits T cell activation by two independent mechanisms. J Exp Med, 198, 1887-1897.

71. Foegeding, N. J., Caston, R. R., McClain, M. S., Ohi, M. D., & Cover, T. L. (2016). An Overview of Helicobacter pylori VacA Toxin Biology. Toxins, 8(6), 173. https://doi.org/10.3390/toxins8060173

72. Yang H, Hu B. Immunological Perspective: Helicobacter pylori Infection and Gastritis. Mediators Inflamm. 2022 Mar 8;2022:2944156. doi: 10.1155/2022/2944156. eCollection 2022

73. Holland RL, Bosi KD, Harpring GH, Luo J, Wallig M, Phillips H, Blanke SR.Chronic in vivo exposure to Helicobacter pylori VacA: Assessing the efficacy of automated and long-term intragastric toxin infusion. Sci Rep. 2020 Jun 9;10(1):9307. doi: 10.1038/s41598-020-65787-3.

74. Salama, N.R., Otto, G., Tompkins, L. and Falkow, S. (2001) Vacuolating cytotoxin of Helicobacter pylori plays a role during colonization in a mouse model of infection. Infect Immun, 69, 730-736.

75. Caso GC, McClain MS, Erwin AL, Truelock MD, Campbell AM, Leasure CS, Nagel M, Schey KL, Lacy DB, Ohi MD, Cover TL. Functional Properties of Oligomeric and Monomeric Forms of Helicobacter pylori VacA Toxin. Infect Immun. 2021 Nov 16;89(12):e0034821. doi: 10.1128/IAI.00348-21.

76. Backert, S. and Selbach, M. (2008) Role of type IV secretion in Helicobacter pylori pathogenesis. Cell Microbiol, 10, 1573-1581.

77. Maeda, S., Ogura, K., Yoshida, H., Kanai, F., Ikenoue, T., Kato, N., Shiratori, Y. and Omata, M. (1998) Major virulence factors, VacA and CagA, are commonly positive in Helicobacter pylori isolates in Japan. Gut, 42, 338-343.

78. Occhialini, A., Marais, A., Urdacı, M., Sierra, R., Munoz, N., Covacci, A. and Mégraud, F. (2001) Composition and gene expression of the cag pathogenicity island in Helicobacter pylori strains isolated from gastric carcinoma and gastritis patients in Costa Rica. Infect Immun, 69, 1902-1908.

79. Noto JM and Richard M. Peek RM. The Helicobacter pylori cag Pathogenicity Island. Methods Mol Biol. 2012; 921: 41-50. doi: 10.1007/978-1-62703-005-2_7

80. Chomvarin C, Namwat W, Chaicumpar K, Mairiang P, Sangchan A, Sripa B, et al. Prevalence of Helicobacter pylori vacA, cagA, cagE, iceA and babA2 genotypes in Thai dyspeptic patients. Int J Infect Dis. 2008;12(5):30-6. doi:10.1016/j.ijid.2007.03.012

81. Mandell, L., Moran, A.P., Cocchiarella, A., Houghton, J., Taylor, N., Fox, J.G., Wang, T.C. and Kurt-Jones, E.A. (2004) Intact gram-negative Helicobacter pylori, Helicobacter felis, and Helicobacter hepaticus bacteria activate innate immunity via toll-like receptor 2 but not toll-like receptor 4. Infect Immun, 72, 6446-6454.

82. Robinson K, Argent RH, Atherton JC. 2007. The inflammatory and immune response to Helicobacter pylori infection. Best Practice & Research Clinical Gastroenterology, 2007, 21(2): 237-259. doi:10.1016/j.bpg.2007.01.001

83. Smith SM (2014). Role of Toll-like receptors in Helicobacter pylori infection and immunity. World J Gastrointest Pathophysiol. 5 (3):133-46.

84. Adamsson J, Ottsjö LS, Lundin SB, Svennerholm AM, Raghavan S. Gastric expression of IL-17A and IFNγ in Helicobacter pylori infected individuals is related to symptoms. Cytokine, 2017, 99: 30-34. doi.org/10.1016/j.cyto.2017.06.013.

85. Wilke CM., Bishop K., Fox D., Zou W. (2011). Deciphering the role of Th17 cells in humandisease. Trends Immunol. 32 (12): 603-611.

86. Akhiani AA, Schon K, Franzen LE et al. Helicobacter pylori-specific antibodies impair the development of gastritis, facilitate bacterial colonization, and counteract resistance against infection. J Immunol 15 Apr 2004; 172(8): 5024-5033.

87. Akhiani AA, Stensson A, Schon K et al. IgA antibodies impair resistance against Helicobacter pylori infection: studies on immune evasion in IL-10-deficient mice. J Immunol 15 Jun 2005; 174(12): 8144-8153.

88. Lehmann FS, Terracciano L, Carena I et al. In situ correlation of cytokine secretion and apoptosis in Helicobacter pylori-associated gastritis. Am J Physiol Gastrointest Liver Physiol Aug 2002; 283(2): G481- G488.

89. Shu X, Ye D, Hu C, Peng K, Zhao H, Li H, Jiang M. Alarming antibiotics resistance of Helicobacter pylori from children in Southeast China over 6 years. Sci Rep. 2022 Oct 22;12(1):17754. doi: 10.1038/s41598-022-21661-y.

90. Dieye Y, Nguer CM, Thiam F, Diouara AAM, Fall C. Recombinant Helicobacter pylori Vaccine Delivery Vehicle: A Promising Tool to Treat Infections and Combat Antimicrobial Resistance. Antibiotics (Basel). 2022 Nov 25;11(12):1701. doi: 10.3390/antibiotics11121701.

91. Pourakbari B, Ghazi M, Mahmoudi S, Mamishi S, Azhdarkosh H, Najafi M, et al, Diagnosis of Helicobacter pylori infection by invasive and noninvasive tests. Braz J Microbiol. 2013; 44(3), 795-8.

92. Gisbert JP, Molina-Infante J, Amador J, Bermejo F, Bujanda L, Calvet X, et al. Fe de errores de "IV Conferencia Espa˜nola de Consenso sobre el tratamiento de la infección por Helicobacter pylori" <[Gastroenterol Hepatol 39 (2016) 697---721]>.Gastroenterol Hepatol. 2017;40:378.

93. Uotani T, Graham DY. Diagnosis of Helicobacter pylori using the rapid urease test. Ann Transl Med 2015;3(1):9.

94. Kim SE, Hwang JH.Management of Helicobacter pylori Infection: A Comparison between Korea and the United States. Gut Liver. 2022 Jul 15;16(4):503-514. doi: 10.5009/gnl210224. Epub 2021 Oct 25.

95. Patel SK, Pratap ChB, Jain AK Gulati AK, Gopal Nath G. Diagnosis of Helicobacter pylori: What should be the gold standard? World J Gastroenterol 2014 September 28; 20(36): 12847-12859.

96. Frickmann H, Zautner AE, Moter A et al. Fluorescence in situ hybridization (FISH) in the microbiological diagnostic routine laboratory: a review. Crit Rev Microbiol 2017;43: 263-93.

97. Mégraud F, Lehours P. Helicobacter pylori detection and antimicrobial susceptibility testing. Clin Microbiol Rev. 2007 Apr;20(2):280-322. doi: 10.1128/CMR.00033-06.

98. Mégraud, F., [When and how does Helicobacter pylori infection occur?].Gastroenterol Clin Biol, 2003. 27(3 Pt 2): p. 374-9.

99. Binmaeil H, Hanafiah A, Mohamed Rose I, Raja Ali RA. Development and Validation of Multiplex uantitative PCR Assay for Detection of Helicobacter pylori and Mutations Conferring Resistance to Clarithromycin and Levofloxacin in Gastric Biopsy. Infect Drug Resist. 2021;14:4129-4145. https://doi.org/10.2147/IDR.S325056.

100. Jordaan M and Laurens JB. Diagnosis of Helicobacter pylori infection with the 13C-urea breath test by means of GC-MS analysis. J Sep Sci 2008;31(2):329-35.

101. Wang S, Zhang WM, Reineks E. Chapter 2. Breath tests for detection of Helicobacter pylori and Aspergillus fumigatus. In : Tang YM and Stratton CW, eds. Advanced techniques in diagnostic microbiology. New York, NY : Springer Science; 2013

102. Mégraud F. Helicobacter pylori infection: good practices. La Presse Médicale. 2010;39,7-8:815-22.

103. Gisbert JP, JM Pajares. 2005. ^{13}C-urea breath test in the management of Helicobacter pylori infection. Digestive and Liver Disease 37 (2005) 899-906.

104. Elbast W and Brazier JL. 1999. Non-dispersive infrared spectroscopy, an alternative to isotope mass spectrometry, the case of the 13C urea breath test. Analusis 1999; 27(3):228-31.

105. Simonova G., and Kalashnikova D. Isotope ratio mass spectrometry application for environmental investigations. E3S Web of Conferences 98, 12020 (2019).

106. Bergman ME, González-Cabanelas D, Wright LP. et al. 2021. Isotope ratio-based quantification of carbon assimilation highlights the role of plastidial isoprenoid precursor availability in photosynthesis. Plant Methods 17, 32 (2021). https://doi.org/10.1186/s13007-021-00731-8.

107. Malfertheiner P, Megraud F, O'Morain CA, Atherton J, Axon AT, Bazzoli F, Gensini GF, Gisbert JP, Graham DY, Rokkas T, El-Omar EM, Kuipers EJ; European Helicobacter Study Group. 2012. Management of Helicobacter pylori infection--the Maastricht IV/ Florence Consensus Report. Gut. 2012 May;61(5):646-64. doi: 10.1136/gutjnl-2012-302084. PMID: 22491499.

108. Gisbert JP, Calvet X. Helicobacter Pylori "Test-and-Treat" Strategy for Management of Dyspepsia: A Comprehensive Review. Clin Transl Gastroenterol. 2013 Mar 28;4(3):e32. doi: 10.1038/ctg.2013.3. PMID: 23535826; PMCID: PMC3616453.

109. Aucher P, Petit ML, Mannant, PR, Pezennec L, Babin P, Fauchère JL. Use of immunoblot assay to define serum antibody patterns associated with Helicobacter pylori infection and with H. pylorirelated ulcers. J Clin Microbiol 1998 (4): 931-936.

110. Widmer M, de Korwin JD, Aucher P, Thiberge JM, Suerbaum S, Labigne A, Fauchère JL. Performance of native and recombinant antigens for diagnosis of Helicobacter pylori infection. Eur J Clin Microbiol Infect Dis 1999 (11): 823-826.

111. Gatta L, Vakil N, Ricci C, Osborn JF, Tampieri A, Perna F, Miglioli M, Vaira D. Effect of proton pump inhibitors and antacid therapy on 13C urea breath tests and stool test for Helicobacter pylori infection. Am J Gastroenterol 2004 (99): 823-239.

112. Murphy G, Freedman ND, Michel A, et al. Prospective study of Helicobacter pylori antigens and gastric noncardia cancer risk in the nutrition intervention trial cohort. International Journal of Cancer. 2015 Oct; 137(8):1938-1946. DOI: 10.1002/ijc.29543. PMID: 25845708; PMCID: PMC4529753.

113. El Hafa F, Wang T, Ndifor VM, Jin G. Association between Helicobacter pylori antibodies determined by multiplex serology and gastric cancer risk: A meta-analysis. Helicobacter. 2022 Jun; 27(3):e12881. DOI: 10.1111/hel.12881. PMID: 35212073.

114. Burucoa C, Delchier JC, Courillon-Mallet A, Korwin JD, Mégraud F, Zerbib F, Raymond J, & Fauchere JL. (2013). Comparative Evaluation of 29 Commercial Helicobacter pylori Serological Kits. Helicobacter, 2013, 18(3):169-79. doi: 10.1111/hel.12030.

115. Sugano K, Tack J, Kuipers EJ, Graham DY, El-Omar EM, Miura S, Haruma K, Asaka M, Uemura N, Malfertheiner P; faculty members of Kyoto Global Consensus Conference. Kyoto global consensus report on Helicobacter pylori gastritis. Gut. 2015 Sep;64(9):1353-67. doi: 10.1136/gutjnl-2015-309252.

116. Graham DY, Lee SY. How to Effectively Use Bismuth Quadruple Therapy: The Good, the Bad, and the Ugly. Gastroenterol Clin North Am. 2015 Sep;44(3):537-63. doi: 10.1016/j.gtc.2015.05.003.

117 Lamarque D., Buruco Ch., Courillon-Mallet A., de Korwin J D., Delchier Ch., Fauchère J L., Kalach N., Labigne A., Lehours Ph., Megraud F., Raymond J. (2012). Revision of French recommendations on the management of Helicobacter pylori infection. Hepato-Gastro & Digestive Oncology, 19 (7) :475-502.

118. Murad S., Said M., Habib A., Elamir K., Mahmoud Amir (2013). Effect of sequential versus standard Helicobacter pylori eradication therapy on the associated iron deficiency anemia in children. Indian Journal of Pharmacology, 45 (5): 470-473.

119. Lamarque D, Raymond J, Megraud F, Moussata D, Heluwaert F, Lehours P, Bazin T, Burucoa C. Comment intégrer a la pratique clinique les recommandations du Groupe d'Etude Français des Helicobacter et de la Haute Autorité de Sante concernant l'éradication de Helicobacter pylori? hepato Gastro 2017; 24: 885-890. doi: 10.1684/hpg.2017.1528

120. Mégraud F. (2004). H pylori antibiotic resistance: prevalence, importance, and advances in testing. Gut 53 (9):1374-1384.

121. Mushtaq A., Rasool N., Riaz M., Tareen RB., Zubair M., Rashid MK., Taufiq-Yape HY., (2013). Antioxidant, antimicrobial studies and characterization of essential oil, fixed oil of Clematis graveolens by GC-MS. Oxid. Commun. 36(4): 1067-1078.

MIX
Papier aus verantwortungsvollen Quellen
Paper from responsible sources
FSC® C105338

Printed by Books on Demand GmbH, Norderstedt / Germany